Jane Blackburn

Birds Drawn From Nature

Jane Blackburn

Birds Drawn From Nature

ISBN/EAN: 9783744651233

Printed in Europe, USA, Canada, Australia, Japan

Cover: Foto ©berggeist007 / pixelio.de

More available books at **www.hansebooks.com**

ADVERTISEMENT.

OPINIONS OF THE PRESS ON PART I.

Recently Published, in Small Quarto, Price 2s 6d.

CAW! CAW! OR THE CHRONICLE OF CROWS.

By R. M., Illustrated in Fourteen Large Page Plates by J. B.

GLASGOW:

PUBLISHED BY JAMES MACLEHOSE, BOOKSELLER TO THE UNIVERSITY.

LONDON: HAMILTON, ADAMS & CO.; LONGMANS & CO.; SIMPKIN & CO.; WHITTAKER & CO.

EDINBURGH: EDMONSTON & DOUGLAS; OLIVER & BOYD; JOHN MENZIES & CO.

DUBLIN: W. H. SMITH & CO.; M'GLASHAN & GILL.

CAMBRIDGE: MACMILLAN & CO.

BIRDS

DRAWN FROM NATURE.

BY

MRS. HUGH BLACKBURN.

GLASGOW:
JAMES MACLEHOSE, BOOKSELLER TO THE UNIVERSITY,
61 ST. VINCENT STREET.

1868.

A

PREFACE.

THE DRAWINGS, of which a few are here engraved, have been made either from the living bird, or from specimens so fresh as to preserve most of the characteristic appearances of life, while the attitude and background have been studied from careful observation of the habits of the wild birds.

This has of course involved a good deal of trouble, and it is not likely that a single observer will have · the opportunity, under these restrictions, of obtaining good drawings of the whole series of British Birds. Such considerations have no doubt induced most illustrators of the subject (even Bewick himself), to put up with a stuffed skin for a lay figure, and, apparently, to label drawings so made as "from nature."

But in the present instance the artist, without neglecting to refer to stuffed "specimens," has refused to be guided by them, in the belief that drawings *really* from nature (and such only) may be made to give a representation of nature more faithful in most essential points than the stuffed skin itself, even when newly set up by the most skilful workmen, and of course in a higher degree preferable to an idealised copy of the usual faded and withered denizen of a glass-case. Even in completeness it is surprising how soon a collection of drawings may be made to bear comparison with all but the greatest museums.

In order to carry out the same idea of interposing as few interpreters as possible between nature and the actual print, the drawings have been copied on to the stone (or zinc plate) by the same hand as made the original drawings, or in some instances the drawing has been made on the stone direct from nature.

After what has been said it will be understood that the choice of subjects has been to some extent limited by circumstances. In many cases, however, it was thought better to give several plates illustrating points of interest in the habits and growth of one species, than to occupy the same space with others, for a complete history of which the materials have not yet been collected. But an untoward accident, which at the eleventh hour has befallen a number of the plates we had intended to include, has destroyed the links connecting some of our illustrations, and made them even less consecutive than they would otherwise have been. Yet it is still hoped that this volume may be considered as so far complete in itself, or at all events as a contribution not without some value towards the illustration of the subject, whether or not opportunity be given to the same hand of continuing the series.

The classificatory names have been copied from Yarrell's excellent work, and a few remarks from personal observation relating to the individual specimen represented, are given, but it has not usually been found necessary to add anything to Yarrell's admirable general descriptions.

It is proper to explain that Roshven, which is given as the locality of some of our birds, is in the district called Moidart, which constitutes the south-west corner of Inverness-shire. It lies on the south shore of the sea loch, Loch Ailort, which is itself a branch of Loch-na-Nuagh, both of which are mentioned.

———————

To the above remarks, which were printed as a preface to the first part of this volume, when published in January, 1862, we have only to add that the plates of the second part (which, though dated 1865, is now first published) are all executed on the same principle,—that the few remarks we thought it worth while to print, have been revised and printed in a more convenient form, but are still restricted to our own personal observations, and that as a contribution to local zoology we have mentioned, if the bird delineated is to be found in the little explored part of the Highlands, which we visit in summer.

GLASGOW COLLEGE, *December*, 1867.

CONTENTS.

I.

HERON'S NEST.

THE details of the Heron's Nest on the Title-page are given from a nest and young which fell from the tree on the very morning we went to ascertain if it could be reached. It was near Killearn in Stirlingshire. The hill in the distance is part of the Campsie range.

BIRDS

Drawn from Nature

By Mrs Hugh Blackburn

Edmonston & Douglas.

Edinburgh, 1862.

II.

SOLAN GOOSE OR GANNET.

THE bird from which the head of the Solan Goose was drawn, was found quite recently dead on the shore near the mouth of Loch Ailort, where we generally see a few in Autumn and in rough weather. We conjecture that this bird, and others which we have found in similar circumstances, had been killed in plunging from a great height into the sea, as these birds do when fishing.

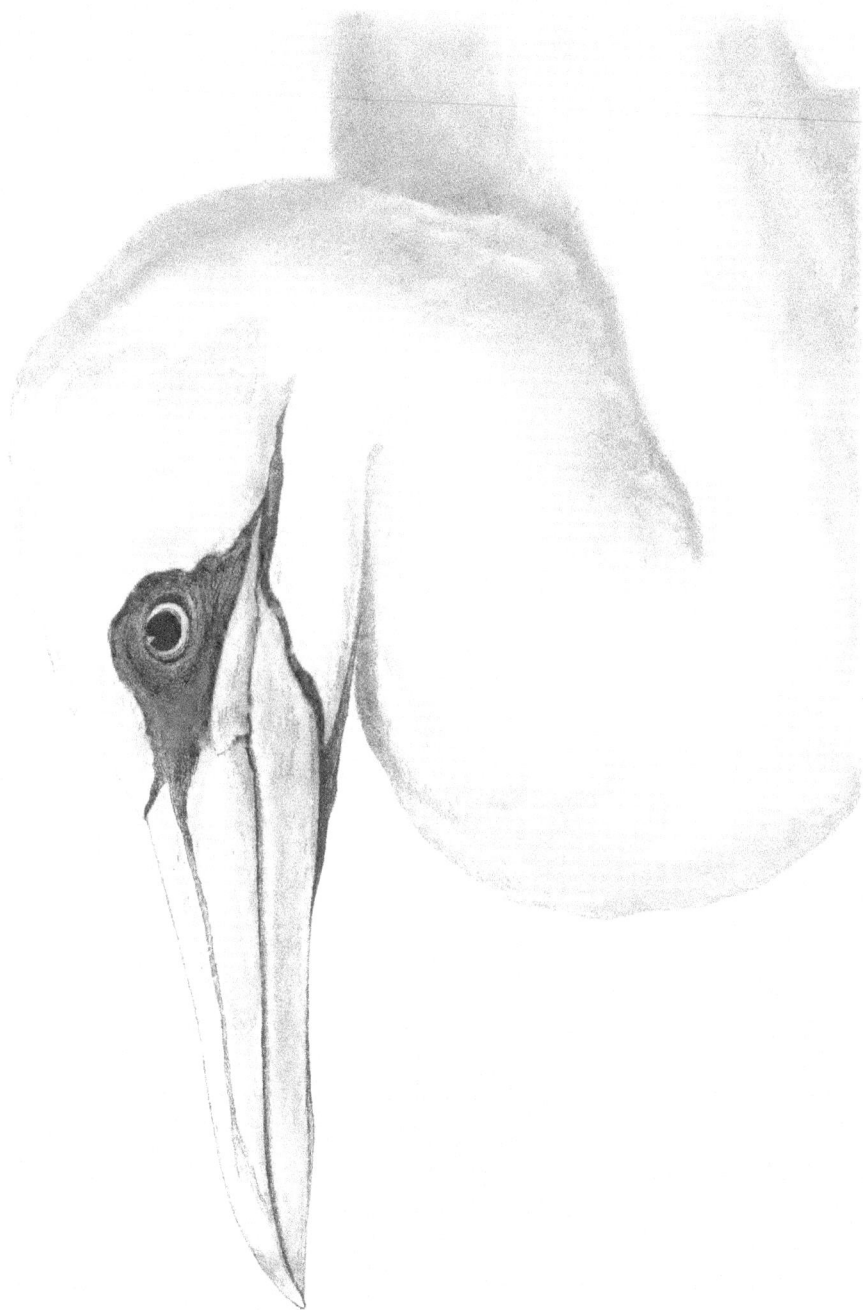

III.

SOLAN GEESE FISHING.

THIS plate is from memoranda of what we have often observed on the Ayrshire coast nearly opposite Ailsa Craig, which is introduced in the distance. On that coast a number of Solan Geese may be seen pouring into a small spot of the sea, and then rising heavily from the water and circling round till they reach a sufficient height, when they dart head foremost down again into nearly the same place.

Bewick's cut of this bird is a great warning not to trust to a stuffed "specimen."

IV

HEADS OF RINGED GUILLEMOT,
COMMON GUILLEMOT, AND YOUNG RAZORBILL.

THE head of the Ringed or Bridled Guillemot was drawn from one shot on 18th April, 1861, on the Clyde at Greenock, by Mr. John Kerr of that town, who states that among a number of Guillemots shot at the time this was the only Ringed Specimen. On the 13th May of the same year we were sufficiently near to one in Loch Ailort to ascertain that it was of this species.

The next head is that of a Common Guillemot in winter plumage from a fresh specimen in Glasgow Market in January.

The third head is that of a Razorbill bought in Glasgow Market in January. It is a bird of the year, and therefore the beak is not fully developed, but the plumage resembles the winter plumage of the adult bird.

Razorbills and Guillemots are common in Loch-na-Nuagh all Summer, but we have not found that they breed there. There are also generally a few Puffins.

The Guillemot flies much better than the small size of the wing indicates. We often see them take long flights, when changing their fishing-ground, in a flock, but have never known one rise from the water to escape. When pursued they (and also the Razorbills and Puffins) invariably dive.

We once had an excellent opportunity of observing the way in which the Guillemot uses its wings under water by letting away, in clear and not deep water, one which had been captured uninjured.

V

THE COMMON GUILLEMOT.

THIS bird was shot in Summer in Loch-na-Nuagh, and sketched at once. It corresponded in all respects with Yarrell's description. The Guillemot in the previous plate, and others in winter plumage in Glasgow Market in January, had the legs and toes orange-yellow, membranes olive and claws black, instead of all black, as in Yarrell's description of the summer plumage of the adult bird. Yarrell does not mention this difference as characteristic of any state of the plumage. It is hard to believe that the legs should change colour twice a-year, and we conjecture from analogy that the yellow colour of the legs belongs to the adult bird : but for this we have no authority. Yarrell also states that the young bird (the plumage of which resembles the winter plumage of the adult) retains the white throat only "till the first spring moult produces the ordinary plumage of summer;" but in Loch-na-Nuagh we have seen at the end of June flocks of Guillemots, some with white and others with black throats and all with black legs. As there is no known breeding place at all near, we suppose that these were all birds of the previous year, driven from the breeding stations, and that many do not assume the black throat till the second year.

VI

THE BLACK GUILLEMOT.

THIS bird was caught alive on her eggs on a little rocky island at the mouth of Loch Ailort, where a few pairs breed annually. She returned to her eggs—one cannot call it her nest—after her picture had been taken.

This bird, unlike the Common Guillemot, readily flies off the water when approached, and does not dive so much. The flight has a resemblance to that of a grouse.

THE BLACK GUILLEMOT,

IN WINTER PLUMAGE.

THIS plate represents the winter plumage of the bird whose summer plumage is given in the previous plate. The specimen figured was shot in Loch-na-Nuagh, in October. This bird is the Dovekie of the Arctic voyagers. It is only in the winter plumage that it has a remote resemblance to a dove.

VIII.

THE BLACK GUILLEMOT—YOUNG AND EGG.

THIS plate is a delineation of the Black Guillemot's nest, or rather of the place where her eggs were laid, on a rocky island in Loch Ailort. The young ones were drawn from life, and were not injured by the operation.

As a correspondent of the *Field* has referred to this plate as a rare instance of the Black Guillemot laying three eggs, it should be stated that two birds had each laid two eggs very near one another in the crevice represented, and as the young birds run as soon as they leave the egg, the birds and egg sketched formed in reality part of two broods.

So far as we have observed the bird always lays two eggs, and no more.

IX

THE SANDPIPER.

WE have found Sandpiper's nests in various positions on the ground, in the side of a bank, or at the foot of a tree, and once only actually in a low bush, but always near the shore. The young birds are able to leave the nest as soon as hatched, but remain near it for a few days, probably returning to it at night.

The nest sketched was in a bank, and the whole family were drawn from life. The old bird was caught by placing the young ones in an open cage, and shutting the door by means of a string as soon as she went in, which she did immediately on our retiring a dozen yards.

There is no perceptible difference of plumage between the male and female, but we assumed this to be the female. They were all restored to their home uninjured.

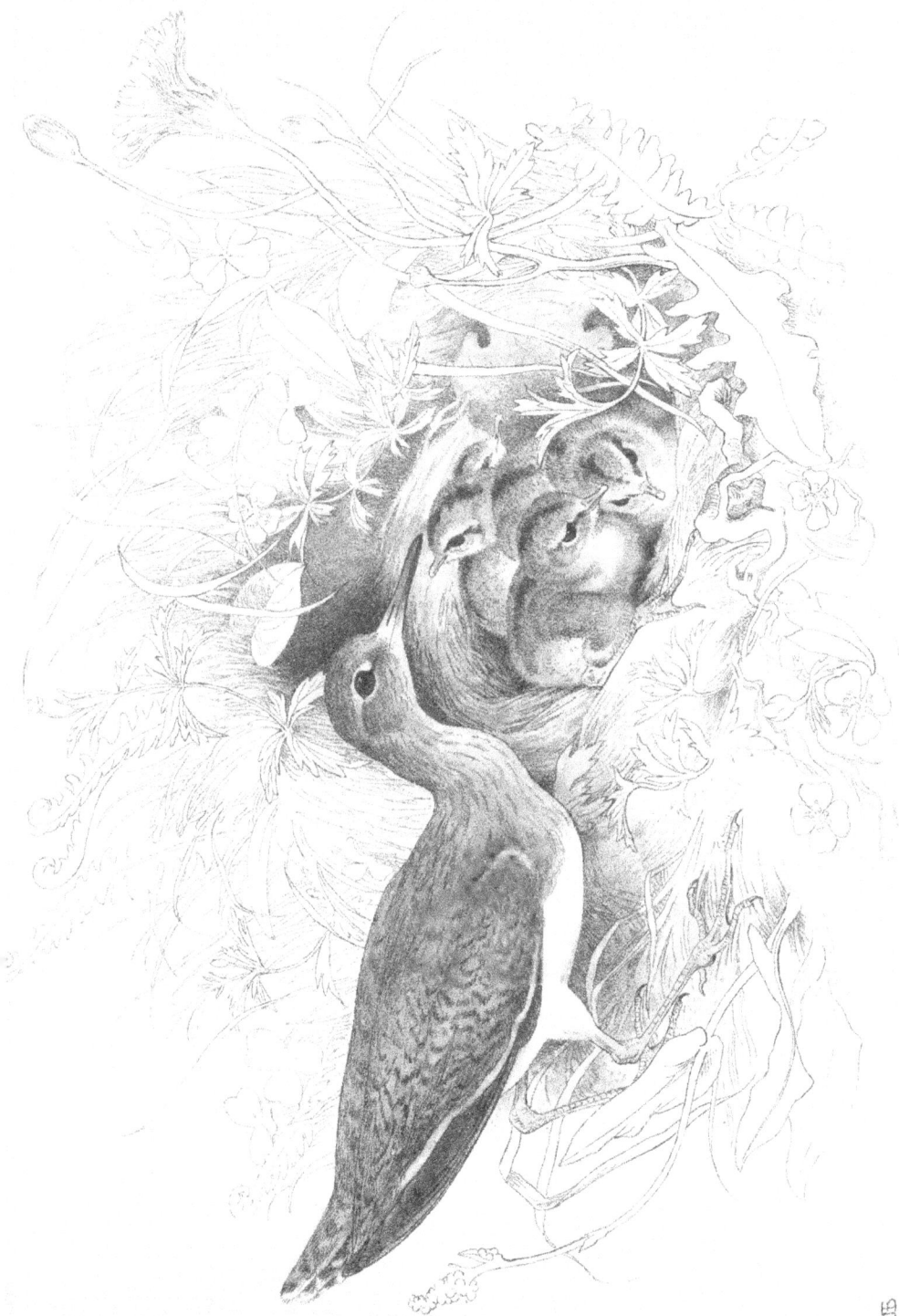

X.

THE FIELDFARE.

THIS plate is from a fresh specimen shot in Forfarshire, December, 1860.

THE SONG THRUSH.

THE drawing of the Song Thrush is from a caged bird. This species is pretty common in Moidart, but hardly so common as the Missel Thrush.

XII.

THE REDWING.

THE Redwing is from a bird caught alive on the snow, near Ayr, in the severe frost at Christmas, 1860.

THE BLACKBIRD.

THIS plate is from a caged bird. The species is common enough in Moidart in gardens.

JE

XIV.

THE BLACKBIRD,

PIED SPECIMEN.

This Pied Specimen of the Common Blackbird was shot at Kinlochmoidart.

THE RING OUSEL.

RING OUSELS breed in considerable numbers on the hills of Moidart. A number of them usually attack the gooseberries in our garden at Roshven when the fruit is ripe. The drawing is from one caught alive after it had been wounded.

The mountain ash berries (or rowans) which the bird is represented eating are a favourite food.

XVI.

THE HEDGE SPARROW.

THIS drawing is from life. The Hedge Sparrow is common in Moidart, but we have never seen the Common House Sparrow there.

XVII

THE WHINCHAT.

THE young Whinchats in the picture are from life; the old birds (male above, female below) are from fresh specimens.

Whinchats are common at Roshven among the whin (furze) which thrives particularly well in some parts of the low ground near the sea.

Wheatears are more numerous, Stonechats not so numerous with us.

THE WILLOW WARBLER.

THE Nest of the Willow Warbler is from nature, and the birds, old and young, from life. The old bird was caught by putting the young ones in a cage. The whole were replaced when the drawing had been made; after which the young were reared by their parents as if nothing had happened. The bird is common in Moidart.

XIX.

THE GOLDEN-CRESTED WREN.

THE Golden-crested Wren (so-called) breeds in Moidart. We have had a beautiful nest suspended by four equidistant pendulous twigs of an old yew-tree: and we have caught the bird alive in the house in autumn.

The twigs were drawn from spruce.

JB

XX

THE BLUE TIT.

THESE birds are drawn from the life. Besides the Blue Tit we have observed the Great Tit or Oxeye, the Cole Tit, and the Long-tailed Tit in Moidart, but no other Paridæ.

The birds are placed on twigs of larch in flower, when the leaf-buds are just bursting, while the last year's cones are still adhering. We mention this as a friendly critic has called the plant hop-bine.

XXI.

THE MOUNTAIN FINCH, OR BRAMBLING.

THIS plate was drawn from a caged bird.

We have not seen the Brambling in Moidart.

B

XXII

THE COMMON HERON.

THE Heron, from which this head was drawn, was caught alive during a hard frost in Stirlingshire. He lived for a fortnight in a garret in the College of Glasgow, where he fished with great success for small herrings in a footpail. He always stood on one leg on a chair without perceptible motion for at least an hour after meals. He seemed quite contented with his lot, but when the mild weather returned he was liberated at the place where he had been captured.

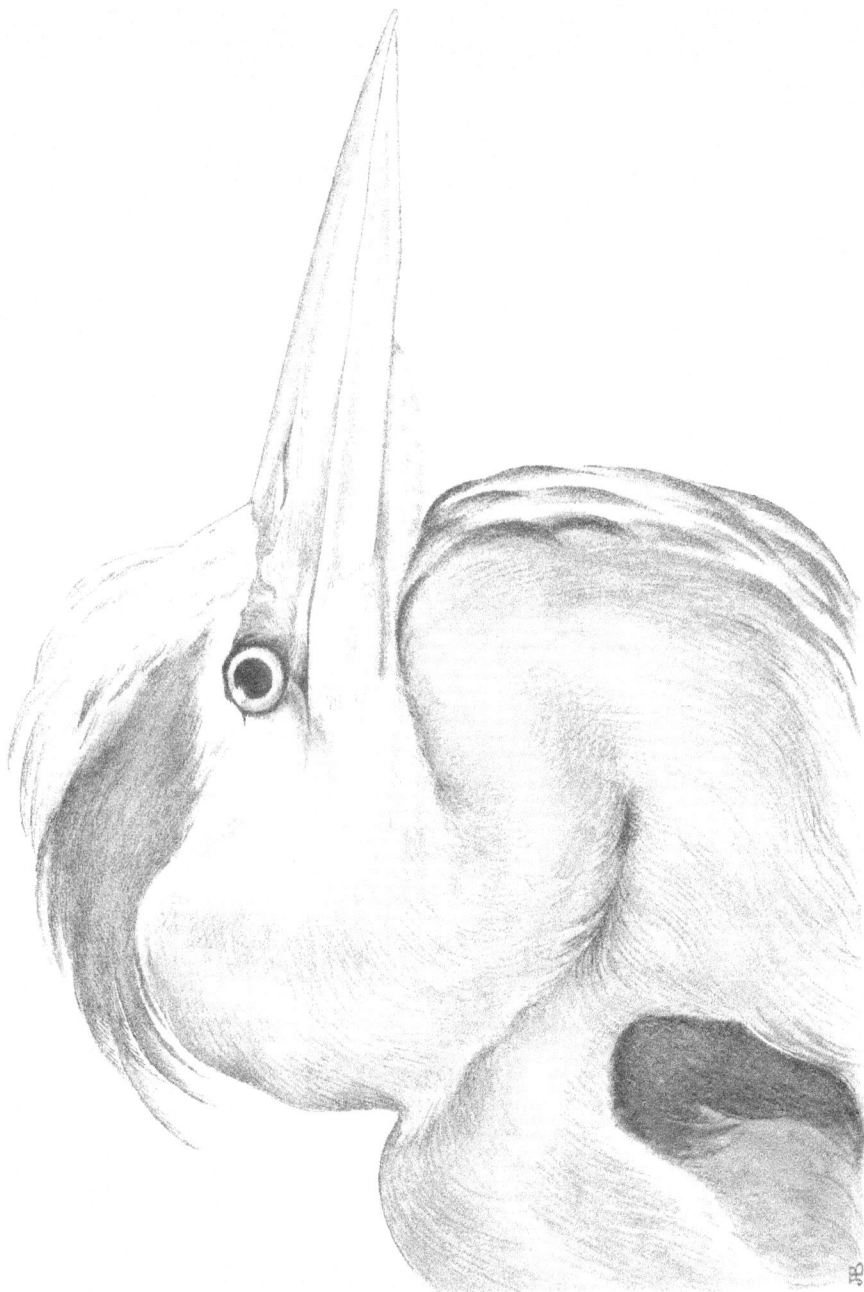

XXIII

GROUP OF HERONS.

THIS Plate represents a group of at least eighteen Herons, seen on Lochiel not far from Fassifern, one fine summer morning. The distant mountain is Ben Nevis.

We have since counted twenty-three Herons at once on the shore of Loch Ailort.

XXIV.

YOUNG GULL.

THE young Gull on the Titlepage to Part II., is from life.

A considerable number of the Herring Gull, and also of the Lesser Blackbacked Gull breed on an island at the mouth of Loch Ailort, where this bird was taken. Terns also breed on this island, as well as on those further up the Loch (more inland), on which the nests of the larger Gulls are replaced by those of the Kittiwake.

The Oyster Catcher breeds on all these islands, as well as both the Redbreasted Merganser *(Mergus Serrator)* and the Goosander or Dundiver *(Mergus Merganser)*.

The Great Blackbacked Gull, which is less plentiful with us, breeds in the fresh water lochs on the hills.

BRITISH BIRDS

BY

Mrs Hugh Blackburn.

PART SECOND,

XXV

GULL'S NEST.

THERE does not seem to be any marked difference between either the eggs or the young of the Herring Gull and of the Lesser Black-backed Gull. The bird in the preceding plate might have belonged to either species—and the same may be said of the nest in the present plate, which, however, we had reason to believe belonged to a pair of Lesser Black-backed Gulls. The drawing was made from an actual specimen *in situ;* and whilst it was in progress the young bird in the egg, which it will be seen is chipped, fairly extricated himself from the shell.

XXVI

THE LESSER BLACK-BACKED GULL.

THE young Gull (whether Herring or Black-backed) represented in the preceding plates, in the course of a few weeks, changes his downy coat for the mottled brown plumage represented to the left of this plate. This plumage is retained for at least one year, and the young birds of the two species, as well as of the Great Black-backed Gull, the plumage of which passes through corresponding changes, are often mistaken for distinct species from their parents. The Wagel or Burgomaster of Bewick, for instance, is the young of one of these three species: just as the Tarrock is the young Kittiwake. The figure to the right is from an adult Lesser Black-backed Gull.

XXVII

GROUP OF SEA BIRDS.

THIS plate represents a very interesting sight which we often enjoy in Loch-na-Nuagh, when the sea birds assemble to feed on herring fry, &c. They then allow a boat to come among them, and one can watch their proceedings. The birds are the Lesser Black-backed and Herring Gull, the Kittiwake (adult, and year old or " Tarrock "), the common Guillemot, and Terns.

The Landscape is in Loch-na-Nuagh, looking westward to the Island of Eigg, over which are seen the hills of Rum.

XXVIII

THE RINGED PLOVER.

THE Ringed Plover, represented in this plate, lays four eggs in a very slight depression of the shingle, without any nest, a little above high water mark. The eggs are not easily found, and the young birds run as soon as hatched,—but the eggs here drawn were watched till the first egg was hatched, so as to have both the eggs and the young in the picture. The birds are common in Loch Ailort (and elsewhere), and collect in Autumn in small flights of twenty or so.

THE RINGED PLOVER
(Charadrius hiaticula)

XXIX

THE WATER RAIL.

THIS drawing is from a freshly killed specimen obtained in the neighbourhood of Glasgow, in January, 1861.

On comparison it was found to differ from Yarrell's description in the following particulars.—The beak, instead of being all red, had the top of the upper mandible nearly black, the rest red, the irides were ↘vermilion instead of hazel, there was a light mark on the under eye-lid ; and some of the . wing coverts (the spurious wing) were slate-grey barred with white, like the flanks. In all these points the bird agreed with Pennant's description and plate.

THE WATER RAIL.

XXX

THE LANDRAIL.

THE Nest of the Landrail, is from nature, at Roshven. The nest was watched till one egg hatched, and the young bird then drawn from life. The next day the whole were hatched, and left the nest. The attitude of the old bird was studied from the life, but some of the details of the plumage were finished from a newly killed bird.

THE LANDRAIL.

(CREX PRATENSIS.)

XXXI.

THE COMMON SNIPE.

THE young Snipe was very carefully drawn from a live bird caught at Roshven in August, and returned safe to its mother. The bird was unquestionably the young of the Common Snipe, and the faithfulness of the portrait may be relied on; but, for some reason or other, it differed much both in form and colour from the plate in Mr. Gould's Birds of Great Britain. The old bird is also from a live specimen, and the background represents the place where the young bird was caught.

THE COMMON SNIPE.
(*Gallinago cœlestis.*)

XXXII.

THE CUCKOO.

THE Cuckoo is drawn from a bird which was caught alive and kept for a few days. There are usually a great many Cuckoos in our neighbourhood, and they are to be seen constantly during the season on the open ground as well as among the trees and brushwood. The attitude represented is not uncommon when the bird settles on the ground in search of insects.

THE CUCKOO.
(*Cuculus Canorus.*)

XXXIII

THE YOUNG CUCKOO.

THIS plate is from an actual example of a Meadow Pipit's Nest on the ground among heather and ferns, which was occupied by a young Cuckoo. The deluded little bird was also observed taking food to her precious foster-child as represented.

THE YOUNG CUCKOO
In Meadow Pipit's Nest
(CUCULUS CANORUS)

YOUNG HOODED CROW.

THE Young Hooded Crow, or Royston Crow, here drawn, was so good as to fall uninjured out of his nest on an old Aspin tree, growing out of a rock on the island at the mouth of Loch Ailort, the day we went to try to get at it.

We have more than once observed the Hooded Crow eating the berries of the rowan (mountain ash) off the trees growing near the sea shore, where there was abundance of shellfish and other apparently more congenial food. His taste for these berries does not seem generally known.

XXXV

CROSSBILLS.

THESE birds were obtained in winter at Ruthven in Forfarshire. The three birds were killed at one shot—they varied considerably in colour.

THE CROSS BILL.

(LOXIA CURVIROSTRA.)

XXXVI.

THE YELLOW AMMER.

THE Yellow Ammer's Nest, among the uncurling ferns and the dog violets, is from nature at Rosliven. The bird remained on her eggs while the drawing was being done.

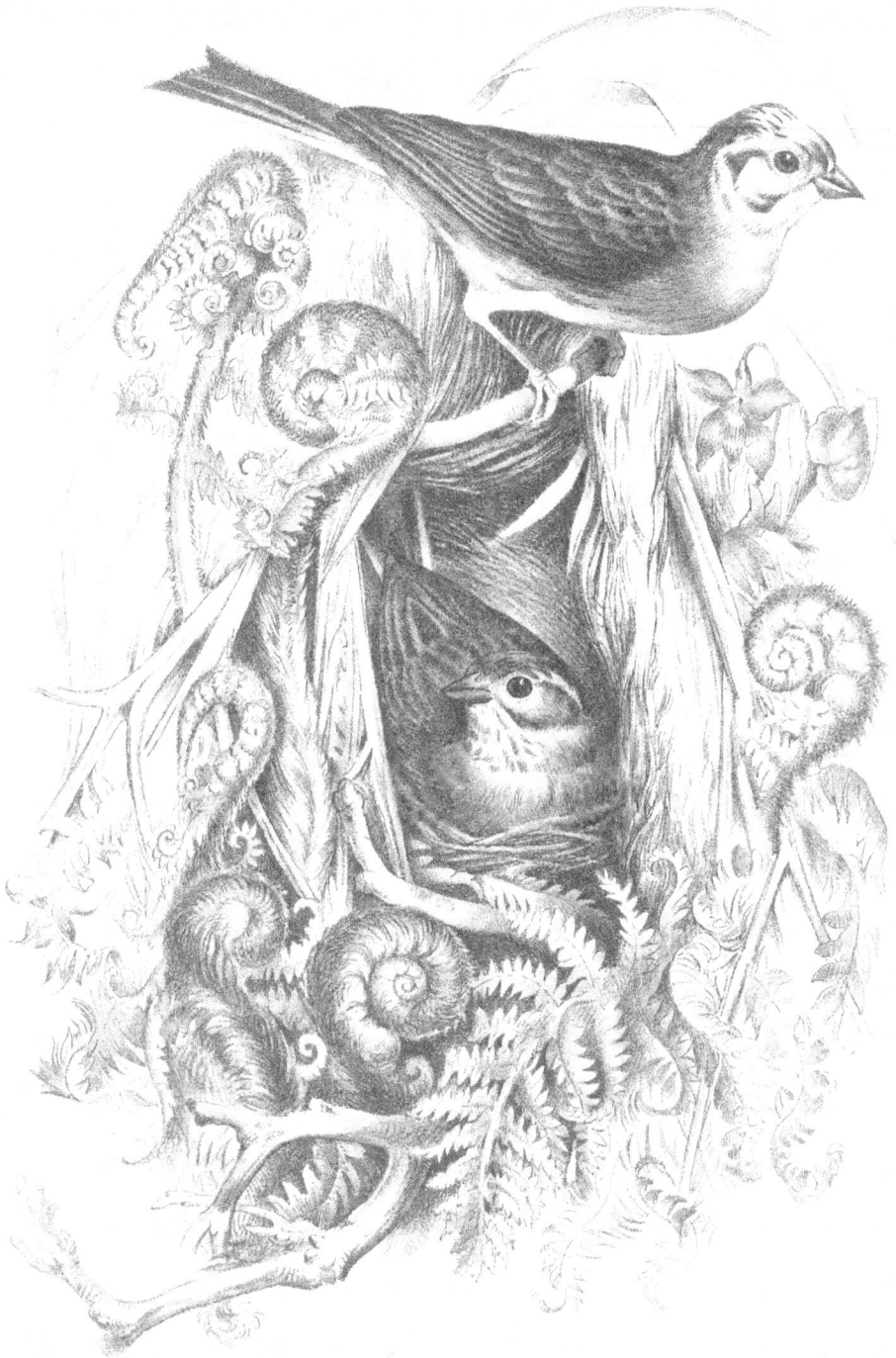

THE YELLOW AMMER.

(EMBERIZA CITRINELLA)

XXXVII.

THE LONGTAILED TIT.

THE Longtailed Tits were obtained at Kinlochmoidart, where they had bred. They had pink eyelids, a point which Yarrell does not mention, but in other respects agreed with his description.

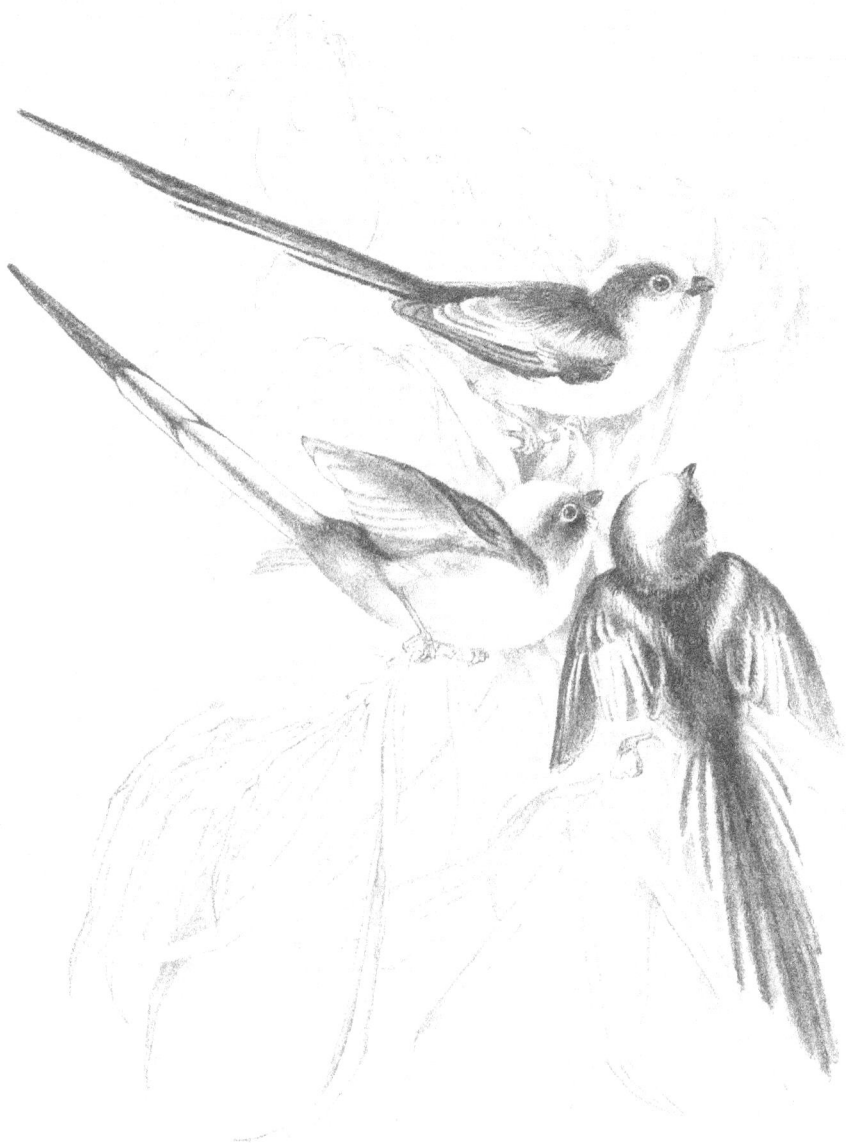

THE LONG TAILED TIT.

(PARUS CAUDATUS)

XXXVIII

THE WOOD WARBLER.

THIS plate is from a fresh specimen shot at Roshven, where it seems to be a regular summer visitor, though Yarrell says he is not aware of any record of its appearance in Scotland. It is, however, less common with us than the Willow Warbler, and than the Common Whitethroat.

THE WOOD WARBLER.
(SYLVIA SYLVICOLA)

XXXIX.

THE STONECHAT.

THE Stonechat is less common in Moidart in summer than either the Whinchat or the Wheatear, but seems to remain all the year—at least we have seen it late in Autumn. The plate represents the male, female, and young bird.

THE STONE CHAT.
(SAXICOLA RUBICOLA)

XL

THE REDSTART.

THESE birds had a nest in our garden wall, in the hole depicted in the plate. We first caught the cock-bird alive, by placing a piece of net over the hole, and when his picture had been taken and he was set at liberty, the hen was caught in the same manner.

THE REGENT.

(Sericulus Chrysocephalus)

THE REDBREAST.

THE upper figure in the plate represents the adult Robin; the lower figure the bird in its first summer. Both figures are from life.

THE REDBREAST.

(*Erythaca Rubecula*)

XLII.

THE TAWNY OWL.

THE Tawny Owl in this plate was a tame bird.

THE TAWNY OWL.
(SYRNIUM STRIDULA.)

THE GOSHAWK.

THE Goshawk was drawn from a trained female bird kept by Mr. Salvin, whose work on hawking is well known. The attitude is, of course, a study from nature.

THE GOSHAWK.

(ASTUR PALUMBARIUS)

HEADS OF GOSHAWK AND KESTRIL.

THE Goshawk's head in this plate is a careful portrait of Mr. Salvin's bird.
The Kestril's head from a fresh specimen is introduced for comparison.

XLV.

YOUNG KESTRILS.

THE Young Kestrils, from which this plate was drawn, were taken alive from the nest, which was on the island at the mouth of Loch Ailort, and the nest was drawn on the spot.

The birds had still a good deal of down, through which the feathers were making their appearance.

YOUNG KESTRELS IN THE NEST.
(FALCO TINNUNCULUS.)